BEAUTIFUL SHELLS
OF NEW ZEALAND

LECTOR HOUSE PUBLIC DOMAIN WORKS

BEAUTIFUL SHELLS OF NEW ZEALAND

EDWARD GEORGE BRITTON MOSS

ISBN: 978-93-5420-183-7

Published: 1908

LECTOR HOUSE LLP
E-MAIL: lectorpublishing@gmail.com

BEAUTIFUL SHELLS
OF NEW ZEALAND.

AN ILLUSTRATED WORK
FOR AMATEUR COLLECTORS
OF NEW ZEALAND MARINE SHELLS
WITH DIRECTIONS FOR COLLECTING
AND CLEANING THEM.

BY

E. G. B. MOSS,

BARRISTER, AUCKLAND.

**PHOTOGRAPHS BY
C. SPENCER, AUCKLAND.**

1908.

PREFACE.

Often have I heard my young friends regret the great difficulty experienced in identifying the things of beauty found on our coast; and some time back it occurred to me that the time had arrived when an attempt should be made to remedy this. New Zealand is a maritime country, most of its inhabitants living near the sea, and there are few indeed who do not enjoy occasionally the pleasure of wandering along the seashore, gathering shells, seaweed, echini, and the numerous other relics of the deep. This pleasant hobby is robbed of a great deal of its interest by a lack of knowledge as regards the names, habits, and mode of preserving the various finds, and especially the finds of shells. When properly preserved and carefully classified they are much more attractive than otherwise they would be. In almost every home shells are seen; some highly prized as ornaments, others as mementoes of pleasant hours in foreign lands; but seldom are our really beautiful shells represented in a collection.

In this work marine shells alone are dealt with, our numerous land and fresh water shells being, with six or seven exceptions, small and insignificant. Of land and fresh water shells about two hundred varieties, and of marine shells about four hundred and fifty varieties, have up to the present been discovered in New Zealand. For some inscrutable reason, however, the New Zealand authorities are continually changing the classical names of our shells. The names I have used are taken from the late Professor F. W. Hutton's last list, published in 1904. It is really time some attempt was made to stop this foolish proceeding. Most of the shells, since I began collecting 20 odd years ago, have had their names changed once, many of them twice, and some even three times. It is more than probable some of the names will be altered while this volume is in the press. These frequent changes in the names cause great confusion, and but for the kindly help and encouragement given me by Mr. T. F. Cheeseman, F.L.S., of Auckland, I should have hesitated to undertake its publication. What most ennobles science is the willingness to give assistance to beginners shown by really scientific men, and doubly pleasing is that help to the recipient when given spontaneously and without stint.

This is the first attempt to publish a popular work on New Zealand shells, and is written by an amateur for amateurs. Nearly every shell likely to be met with by an ordinary collector (except the minute shells) will be found in the ten plates at the end of this work. I have endeavoured to describe the shells in simple language, as the scientific words may puzzle some of my readers. For instance, Professor Hutton describes a certain shell as "thick, irregular, sharp ribbed, with the margin dentated or lobed, very inequivalve; upper valve opercular, compressed, wrinkled, with thick concentric laminae; lower valve cucullated, purple, white within, edged with purple or black; lateral margins denticulated; hinge generally attenuated, produced, pointed." When a shell is found that fully answers this description you will know it is an Auckland rock oyster. Errors and omissions will, I trust, be charitably dealt with, as the inevitable mistakes of a man who is blazing a track. I have endeavoured to give the Maori names also, but, unfortunately, in different parts of New Zealand the same name is frequently used for different shells.

My own collection of New Zealand marine shells, made during my residence in Tauranga, Bay of Plenty, is, I believe, the best and largest yet made, and among the specimens I can number no less than a dozen new shells which I had the pleasure of adding to the recognised list. Over 90 per cent. of the known species of New Zealand marine shells were found there by my friends or myself during the 15 happy years I spent in that delightful, though not very progressive, part of New Zealand.

My thanks are especially due to Mr. Charles Spencer, of Auckland, an ardent conchologist, and for many years my colleague in collecting shells, for the care taken with the photographs, and for valuable suggestions and help.

CONTENTS

CHAPTER I.

SHELLS AND THEIR INMATES.

Before the study of shellfish, or molluscs, was conducted on the scientific principles of the present day, shells were classified as univalves, bivalves, and multivalves. The univalves were shells in one piece, such as the whelk; the bivalves those in two pieces, such as the mussel or oyster; and the multivalves those in more than two pieces, such as barnacles or chitons, barnacles, however, being no longer classed with shells.

The highest of the five types, or natural divisions, of animals are the Vertebrata, the Mollusca, and the Annulosa. The vertebrates usually have vertebrae, or jointed backbones, and from this the highest division takes its name; but the real test is the colour of the blood, which in the vertebrates is always red.

The molluscs have soft bodies and no internal skeleton, but in lieu of this the animal is usually protected by an external shell, harder than the bones of vertebrates. The annulosa, like the molluscs, have soft bodies and no internal skeletons; but the external shell is divided into joints or segments, and is usually softer than the bones of vertebrates.

Fishes belong to the vertebrate division, oysters to the mollusc, and crabs and starfish to the annulosa.

The remaining two of the five divisions are the Caelenterata, in which the general cavity of the body communicates freely with that of the digestive apparatus, and the Protozoa, which includes all animals, such as sponges, etc., not included in the above four divisions.

The shell of an oyster takes the place of the bones of a dog; and although it may seem strange for an animal to have its bones on the outside of its body, it is really no more strange than for a fruit, such as the strawberry or raspberry, to have its seeds on the outside. Lime is the principal ingredient of all bones; and the bones of vertebrate animals contain a large proportion of phosphate of lime, while the shells of molluscs, or shellfish (as they are popularly called), consist almost entirely of carbonate of lime.

When scientists began more carefully to examine the structure of shellfish, they found that those similarly constructed had shells with certain marked peculiarities. The days of conchology were then doomed; and the study of the mollusc, or malacology, took its place.

Besides those necessary for digesting food, most shellfish have organs equivalent to those of vertebrate animals, such as feet, arms, eyes, head, heart, and tongue. Although bearing the same names, these organs rarely have a similar shape to those of the vertebrates, being necessarily adapted to the different mode of living. The foot of a cockle, shaped like an animal's tongue, enables it to move slowly from place to place, as well as to burrow in a sandy beach with the comical jerks so well known to observers. The tongues are beautifully designed for their work. The long, narrow tongue of the vegetarian mollusc works like a scythe, and mows down the delicate marine grasses on which the animal feeds. The powerful tongues of those that prefer an animal diet are able to bore through the strongest shells; and woe betide the unfortunate shellfish which, having shown signs of weakness, or disease, is surrounded by its active, carnivorous brethren. The tongue, sometimes longer even than the shell itself, is covered with rows of very hard spikes, or teeth, arranged similarly to the burrs on a file. As these teeth break, or are worn out, they are replaced by others that push themselves forward when wanted. Under a microscope of moderate power, the radula, or tongue, of a shellfish, especially a limpet, is a most interesting sight, and many molluscs can be identified merely by examining the tongue under a microscope. The shape of the teeth, the number, and the arrangement of them will settle the question.

The appetites of molluscs verge on the voracious. Break up a few cockles, or other shellfish, and place them in shallow water on a calm day, and watch the result. If in the vicinity of rocks, and during a rising tide, all the better. First come the wary little shrimps to the feast. Some are creeping cautiously, and some are jumping and racing, as if afraid of not being in time. Then the carnivorous shellfish approach from all directions, foremost amongst them being the different species of Cominella. While they are lumbering along, shells appear to be actually running; but a close inspection shows that these contain active little hermit crabs, whose tender tails, having no hard covering of their own, are snugly stowed in the empty shells of defunct molluscs. Then the sand or gravel moves, and crabs appear. The shrimps, crabs, and hermit crabs run off with the smaller morsels; but the molluscs gather round the remnants and pull and haul and roll over one another until the feast is ended, when some, being satiated, contentedly burrow into the sand; while others, with their appetites only sharpened, will wander away in search of fresh prey.

In many shells, such as the Triton, or Lotorium as it is now called (Plate III.), every increase in growth can be traced in the thick lip formed by the animal when it has increased the size of its shell. Others again, such as the Struthiolaria (Plate IV., Fig. 4), only form a lip when their full size has been attained, and by this the difference between an old and young Struthiolaria can at a glance be seen. Others form a lip at each growth, and then dissolve the lip before starting again. Vertebrate fish are supposed to grow, and increase in size, till the day of their death, but shellfish do not do this. The shell becomes stronger and thicker with age, the animal having the ability to add layer after layer of nacreous, or pearly deposit, on the inside of the shell; and as the animal shrivels and lessens in size the thickness of the shell increases. And some, when they become too large, have power to dissolve the partitions in the shell, and deposit the material on the outside of the shell.

The time it takes a shellfish to grow to its full size varies a great deal. Oysters take about five years; but the giant Tridacna, the largest bivalve in the world, has been found so enclosed in the slow-growing coral that it could hardly open its valves.

The young of most shellfish are active little things, and are usually so different from their parents as to be unrecognisable. Some swim, or frisk about, and travel even long distances in search of suitable quarters to settle in. Others float on the surface, and are driven where the winds and currents list. Some, like mussels, are distributed all over the world, others again are found, perhaps, on one rock, or on one small sandbank in a large district. Many shells are rare, because we do not know where to look for them; but if we know and can find their food, we will find the shellfish not far away. Some change their shape so much that, as they age, they have to dissolve all the partitions made in their youth in the shell. The eggs of some are scattered on the surface of the water, while the eggs of others are hatched by the mother before being turned adrift.

Marine shellfish live in all kinds of places below high water mark; and some of the semi-amphibious ones thrive even above ordinary high water mark, where for days at a time nothing but the tops of the waves could reach them. They are found on seaweed and on rocks, and on sand or mud-banks; but especially in places near rocks on marine grass banks bare at low spring tides. Some live on the surface of the water, some burrow in sand or mud, and some bore holes for themselves in the softer rocks. Some live in deep water; but the better coloured shells are found near low water mark, or in shallow water; for light is as necessary to the perfecting of colour in shells as in flowers. Shells that have grown in a harbour are more fragile than those grown in the ocean, and are usually less brilliant in colour, as harbour water is not as clean as ocean water. The colour of shells (as of insects) depends largely on environment, and is only one, and by no means the most reliable, method of deciding the species. An expert can at a glance tell whether a given shell has come from shallow or deep water, and whether from an exposed or sheltered spot. Most shellfish move about a great deal, and migrate into deeper water in summer; and on bright clear days retire into dark corners amongst, and even under, stones. On a dull day a collector is frequently more successful than on a bright, sunny day; and in spring or early summer the best hauls of live shells can be made. Nearly all shells have an epidermis, or outer skin. In some this is very apparent, as in the Lotorium olearium (Plate V., Fig. 1), or the Solenomya parkinsoni (Plate IX, Fig. 18), while in others it is nearly transparent, and hardly perceptible. To enable the true colours of a shell to be seen the epidermis must be removed.

The supposed original form of a shell was that of a volute univalve, such as the Triton (now Lotorium), or Struthiolaria. To properly enclose the animal, and make it safe from enemies, an operculum, or lid, was so formed that when the animal retired into the shell this filled up the opening. The operculum is usually like a piece of thin, rough brown horn, and where no reference is made to an operculum in this work, it must be understood that the operculum is horny. Some shells, such as the Astralium sulcatum (Plate VI., Fig. 18), and the Turbo helicinus (Plate VI., Fig. 17), have a shelly operculum; that of the latter being the well-known cat's eye.

In some shells the operculum is small, in others large, and progressing step by step we find some, such as the scallop and oyster, with one side round, and the other (really an operculum) flat and as large as the shell; until we come to the perfect type with each valve the same shape and size. Then the operculum disappears, as in the limpet, and the covering shell becomes smaller and smaller, till in the Scutum ambiguum (Plate IX., Fig. 23) the shell bears about the same proportion to the animal that the little bonnet, fashionable a few years ago, bore to the lady that wore it. The shell is built up of very thin layers of nacre, or mother of pearl, and calcareous or chalky matter, the thinner being the layers of nacre the more lustrous and iridescent is the shell.

As would be expected from its isolated position, many of the genera of New Zealand shells are not found elsewhere. The late Professor Hutton mentions nine genera in this position.

The dispersal of shells is an interesting natural phenomenon. The eggs of molluscs are so small that they can easily be carried by currents, attached to floating seaweed or floating timber, on the hulls of ships, or in the feathers or feet of our migratory birds, such as the godwit, which every year travels from New Zealand to Siberia and back. A great many of our shells are found on the Australian coasts; and a surprising number are common to both New Zealand and Queensland.

In describing the illustrations, length means extreme length, and by measuring the shell on the plate the proportionate width can be ascertained. The illustrations are, generally speaking, half the natural length of the shell depicted; and the shell photographed, although in most cases an average full-sized specimen, in some instances was smaller than the average.

CHAPTER II.

COLLECTING AND CLEANING SHELLS.

Shells are described as live and dead shells. Live shells are those found with the animal enclosed, and are more likely to be perfect in form and colour than dead shells. Dead shells found amongst rocks are nearly always battered and worn, and useless from the collector's point of view. Live shells are found below high water mark, among rocks, or in the sand, or amongst seaweed and marine grasses.

Wait till a storm from the sea is ended, and then, if the wind is blowing from the land, a rich harvest of live and dead shells will be found on the sandy beaches and amongst the seaweed and wrack that comes ashore. Many of the smaller shells will be found amongst the leaves and roots of kelp. Start early in the morning, or pigs, rats, and seabirds will have destroyed the choicest specimens. Even such solid bivalves as the Dosinia will be carried skywards by the gulls and dropped on to a hard part of the beach, so that the shells may be cracked and the gulls get the contents. Most birds have this habit; even thrushes can be seen carrying snails up in the air and dropping them on to paths. Soak the dead shells in hot water for a few hours to get rid of the salt, and then scrub with a hard brush, or, if encrusted or very dirty, rub with sand, using a brush or cloth. No need to fear hurting them, unless very fragile, in which case the best thing is a soft toothbrush, with fine sand. If patches of dirt, or encrustations, still remain, scrape with a piece of hard wood or a knife. As a last resource use muriatic acid, diluted with an equal volume of water; but be careful to put it only on the spots to be cleaned, using a penholder, or small stick, with a small piece of rag tied to the point. The inside of the shell, if discoloured, can be cleaned in the same way. When cleaned, wash again carefully, and dry thoroughly. Then rub the shell with a mixture of sewing machine oil and chloroform in equal parts. The machine oil, being fish oil, will replace the oil the shell has lost, and chloroform is the best restorer of colour we have. For very delicate shells poppy oil is sometimes used; but it is expensive and difficult to obtain.

The greatest trouble is getting the animal out of live shells. Anthills are few and small in New Zealand, so the lazy man's method of putting shells on an anthill, and letting the insects do the work, is impracticable. Boiling for a minute will not hurt the stronger and heavier shells; but even pouring boiling water on the more delicate shells will cause them in time to fade. After taking the shells out of the boiling water, let them cool, and then place them in cold, fresh water for a couple of days in summer or for a week in winter, changing the water every day. The animal can then usually be removed with a bradawl, or, better still, a sail needle stuck into a cork. Although soaking in fresh water for a few days makes the animal slip out more easily, still a large proportion will break during extraction. The piece left behind must also be extracted, or the shell will be offensive. The coarser shells can be buried for a few months in sandy soil, or for a few weeks on a sandy beach below high water mark, or put in baskets or bags made of twine or netting, and placed in tidal pools, or fastened to stakes at low water mark, where the marine insects will quickly do their share of the work. Or they may be buried in a boxful of clean sand or sandy soil, and the sand kept moist by watering it every few days. The box is all the better for being put away in a damp place under a tree, or on the shady side of a building or fence. This, however, is a slow process, and if the specimens are required at once, the best way is to extract all you can of the animal by the hot water and soaking process, and then keep the shell half-full of water in a shady place, every morning holding it under a water tap and shaking it carefully. After each shaking a very little pure muriatic acid may be put into the shell, and when all the effervescing from the acid is over, wash and shake it again. Two or three mornings of this treatment should clean the shell. The more delicate shells will lose their colour if put into boiling water, so first put the boiling water in a basin and then place the shells in it. Nearly all salt water shellfish, if soaked for a few hours in fresh water, will die. The only exceptions I know of are the Nerita and Littorina, families which are semi-amphibious. The best way to remove coral or vegetable growths from shells is to leave them for a few weeks, or if very hard, for a few months, in a shady place, where the wind and rain can get at them, but not the sun. The growths will then be sufficiently soft to be scraped off with a piece of hard wood or a knife, or rubbed off with sand. It is a good plan to oil or paste calico over portions not covered with growths, so as to reduce the risk of the colour fading. When the animal is removed and the growth cleaned away, wash, scrub, and dry, as with dead shells.

Shellfish are sometimes obtained by dredging with a naturalist's dredge, or by diving for them, or lifting them out of the water with instruments such as hay forks and hooks. Sandy beaches and banks yield many of the most beautiful specimens, but only with experience will the collector be able to identify the marks of the syphons of the various shellfish. Nearly all shellfish that burrow have two syphons, or tubes, which they push through the sand. The water is drawn down one syphon and up the other; and as it passes through its stomach the mollusc absorbs the animal and vegetable particles in the water. Some of these shellfish live feet below the surface of the sand; some, such as the common cockle, only a fraction of an inch. Apparently even cockles do not come to the surface, except to die. Some instinct seems to urge a shellfish, when sick unto death, to save its fellows from infection by leaving the common shelter. Cockles found on the surface are to be avoided as unhealthy, and, unless they die naturally, are soon killed by the carnivorous shellfish. It does not take one of the whelk family long to bore a hole in the centre of the cockle shell. It knows too much to risk having its radula, or tongue, nipped off by putting it between the partly-open valves of the dying cockle. The end of the syphon, which projects from the sand, is like a miniature sea anemone. Each sand-burrowing shellfish has a different shaped end to its syphon, and the skilled collector can tell at a glance what shellfish is down below. If he can grip the syphon with his hand he will have no difficulty in digging up the shellfish, even such a deep-living one as the Panopaea (Plate VIII., Fig. 3), one of which was captured by Mr. C. Spencer on Cheltenham Beach, near Takapuna Head, in Auckland Harbour. I believe this was the only Panopaea captured in New Zealand in situ, and was about eighteen inches below the surface of the sand at half-tide mark. If he miss gripping the syphon he will probably lose the shellfish; as it can burrow nearly as fast as a man can dig with his hand. A beginner cannot do better than take a small spade, and walk along a sandy beach at low water. As the tide begins to rise, and the buried shellfish feel the water, he will see the sand moving, or showing signs of life; and if he digs quickly enough he may unearth rare and beautiful specimens for his cabinet.

Wherever animals or vegetables are crowded, disease appears. This is true of molluscs, and it is seldom worth while looking for a specimen fit for a collection where any particular kind of shellfish lives in great numbers. Animal and vegetable parasites will be found wherever shellfish are crowded together. For instance, a perfect cockle, or one good enough for a collection, will not be found on a cockle bank, but solitary ones must be looked for elsewhere.

CHAPTER III.

DESCRIPTION OF PLATES.

Amongst the best known shells in any part of the world the Nautilus takes a leading position. Named Argonauta by scientific men, after the Argonautae, or sailors of the Argo, it has been the subject of many legends from the earliest times. Aristotle describes it as floating on the surface of the sea in fine weather, and holding out its sail-shaped arms to the breeze. This is now known to be incorrect, as the use it makes of these arms is to help it in swimming through the water. New Zealand's specimen, the Argonauta nodosa, also known as Argonauta argo, the most beautiful of the four known species, is depicted on Plate I. Being a floating shell, and found even hundreds of miles from land, our Nautilus is not peculiar to New Zealand. Its beautiful white, horny-looking shell can be obtained from most parts of the Pacific and Indian Oceans, but in no part of the world can finer specimens than ours be found. It is known to the Maoris as Muheke or Ngu, and colloquially as the Paper Nautilus.

The animal that produces this shell belongs to the octopus, or cuttlefish, family. The male is an insignificant-looking octopus, about an inch long. The female grows many times larger, as can be imagined from a glance at the shell in the plate, which measured nine inches across, and was found at Mayor Island, in the Bay of Plenty, and is now in the possession of Mr. C. Spencer.

In the shell the female lays her eggs, and in it the young are hatched. Unlike all other shells, the Nautilus is not moulded on the animal, nor is she even attached to her shell by muscles. When washed ashore she can wriggle out of her shell and swim away. In her shell she lies as in a boat, propelling herself by slowly sucking up water, and violently ejecting it through a funnel, or syphon, at the same time using her arms as oars, to increase her speed. Dame Nautilus can sink to the bottom of the sea if she chooses; and when wishing to crawl about the sand or rocks she turns over and carries her shell on her back, like a snail.

Beside the Nautilus is her little cousin, the Spirula peroni, which sometimes, although not quite scientifically correct, is called an Ammonite. Our Nautilus is frequently found alive, but only one living specimen of this Ammonite has hitherto been caught, though several shells have been obtained from different parts of the world with portions of the fish attached.

Neither towing nets nor dredges have been successful in catching the Ammonite, so it evidently does not live either on the surface or bottom of the sea, but probably between the two, in deep water. The shell is in a number of divisions, connected by a fine tube, and no doubt its use is to regulate the depth at which the animal wishes to stay. This the creature does by filling a number of the divisions with water or air, according as it wishes to sink deeper or float upwards. After a gale, on looking amongst the wrack cast up by the highest waves, large numbers of our Spirula will be found. Light and fragile the shells are, and they ride ashore without injury, and frequently are found covered with small barnacles, a proof that many weeks must have elapsed between the death of the owner and the casting ashore of its shell. In places in New Zealand, and elsewhere, large fossil deposits of Spirula peroni occur. It is worth remembering that, even though this shell is found as far away as England, the only living specimen was caught on the New Zealand coast. Our only other floating shells are three species of Janthina, or violet shells, two of which are shown on Plate VII., Figures 1 and 2.

The first three shells on Plate II. belong to the Murex family. From this species the ancient Tyrians obtained a portion of their celebrated purple dye. The Janthina family (Plate VII.), however, contributed the greater portion. The dye was extracted by bruising the smaller shells in mortars.

MUREX ZELANDICUS (Plate II.).—Fig. 1 is known as the spider shell, from the spines, which look like spider's legs. It is a white or greyish shell, about two inches in length. The long spines would interfere with the growth of this Murex if it had not the power of dissolving them as the outside of one whorl becomes the inside of the next. The removal is supposed to be assisted by chemical action, as the saliva of some shellfish is known to contain a small percentage of muriatic acid. Such powers have some shellfish of dissolving or altering the form of their shells, that the Cyprae, or Cowry, our representative of which family is the Trivia australis (Plate VII., Fig. 29), not only can dissolve the inner part of its shell, but can deposit new layers on the outside. This Murex lives on sand in the open ocean, and is found in the North Island only.

MUREX OCTOGONUS (Plate II.).—Fig 2 is a slightly longer shell than the Murex zelandicus, and, like it, is found only in the North Island. But in place of being round or oval, this shell is octagonal, from which peculiarity it derives its name. The grooves that cross the shell are deep, and between them are small curved spines. The shell is thick and solid, the exterior being reddish white, sometimes stained with brown. There is a smaller variety of this shell, darker in colour and with more numerous spines than the photographed specimens shown.

MUREX EOS (Plate II.).—Fig. 3 is a beautiful pink shell, about an inch long. Dead shells only have been found, and a good specimen is much prized. None of the Murex family are common, and they are seldom found alive. Murex eos, although existing in Tasmania and Australia, has so far been found in New Zealand nowhere South of the Bay of Islands.

MUREX RAMOSUS.—Two specimens of this well-known Island shell have been found in Tauranga during the last five years. One excellent specimen, 8½ inches long, was a live shell, and is now in the possession of Mrs. T. M. Humphreys, of Tauranga. An illustration of this shell will be found on Plate X., Fig. 10.

TROPHON STANGERI (Plate II.).—Fig. 4 is a rough grey shell, with a dark purple interior. It is covered with parallel ridges and lines, which are known as varices, very thin and close together, and running from the apex to the mouth of the shell. It is over an inch in length, and usually found on cockle banks in harbours.

TROPHON AMBIGUUS (Plate II.).—Fig. 5 is in shape very like the Murex stangeri, but twice the dimensions, and can be easily distinguished, as the varices are much higher and further apart; besides which they cross one another at right angles, forming a perfect network, and the interior is pinkish brown. This shell is found on ocean beaches, as well as on cockle banks.

TROPHON CHEESEMANI (Plate II.).—Fig. 6 is a small, grey Trophon, with a dark interior. The shell is deeply grooved, and about three-quarters of an inch long. Found, so far, only on the West Coast, near Waikato Heads. We have 3 other small Trophons, two of which are shown on Plate VII., Figs. 22 and 23.

ANCILLA AUSTRALIS (Plate II.)—Fig. 7 (also known as the New Zealand Olive) is a beautiful clean bright shell, and looks as if covered with shining enamel. The upper part of shells of the Ancilla family is kept polished by the mollusc's foot, which swells to such an extent when the animal is moving about that the whole shell is concealed in its folds. The broad band in the centre is usually dark chestnut or brownish purple, the points of the shell being tipped with darker shades of the same colour. The interior is purplish. Large numbers are found on the edges of channels in harbours, buried in the sand; but their presence is easily located by the oval-shaped mound under which they conceal themselves. When washed up on ocean beaches, they are frequently bleached to a brown or chocolate colour. The Maoris sometimes use them for buttons, and very pretty buttons the medium-sized ones make. The largest I have seen were two inches long. There are two other kinds of Ancilla found in New Zealand, the one much larger, and the other much smaller, than the one depicted. The larger is Ancilla pyramidalis, the smaller Ancilla mucronata. The native names are Pupurore and Tikoaka.

PURPURA SUCCINCTA (Plate II.).—Figs. 8 and 9 is found all over the North Island, on ocean beaches and in harbours. It may have a comparatively smooth exterior, as in Fig. 8, or be deeply grooved, as in Fig. 9. The interior is usually yellow or brown, and generally has a pale band round the margin of the outer lip. It is very variable in colour and general outside appearance, and although at one time divided by naturalists into 3 or 4 varieties, under different names, it is now believed to be only one very variable species.

PURPURA SCOBINA (Plate II.)—Fig. 10 (late Polytropa scobina) is a rough, thick, brown shell, with a dark interior. It varies in colour and shape, and is found everywhere in New Zealand on surf-beaten rocks. It is usually under an inch in length.

PURPURA HAUSTRUM (Plate II.).—Fig 11 (late Polytropa haustrum) is a brown shell, with a greyish or yellow interior. It is found in great numbers on rocks in all parts of New Zealand. Sometimes it is over three inches in length. The animal equals the Cominella in voracity. The Maori name is Kakare, or Kaeo, both of which names are also given to the Astralium sulcatum (Plate VI., Fig. 18).

SCAPHELLA PACIFICA (Plate II.).—Fig. 12 (late Voluta pacifica) is a yellow or chestnut-coloured shell, with dark markings, and is sometimes nine inches in length. It is found in large numbers washed up on the beaches in both Islands after gales, and varies so much in colour, markings, and shape that a good pair is seldom procurable. Sometimes even the nodules, or lumps, shown in the plate, are wanting, and sometimes the markings are wanting. It was until lately known as the Voluta pacifica, being one of the well-known Volute family. It lives in the sand on exposed beaches. The Maori name is Pupurore, which name is also used for the Ancilla australis (Plate II., Fig. 7).

SCAPHELLA GRACILIS (Plate II.).—Fig. 13 (late Voluta gracilis), besides being smaller and narrower than the Scaphella pacifica, is distinguished by the markings, which in the latter appear to form bands, while in the former they do not. With such a variable shell, however, it is difficult to distinguish the one from the other.

MITRA MELANIANA (Plate II.).—Fig. 14 is a dark chocolate-coloured mitre-shaped shell. Being smooth and of the same colour, both internally and externally, it cannot be mistaken. About a score of dead ones, varying from one and a-half to two inches in length, have been found by my friends and myself on the ocean beaches near the entrance to Tauranga Harbour, and at Maketu, in the Bay of Plenty. This is a particularly interesting discovery, as the Mitre shells (so called from their shape resembling that of a bishop's mitre) hitherto found out of the tropics were minute. We have one other Mitre shell, which is pink or brownish, and under one-third of an inch long.

Plate III. represents two of our largest and most handsome shells. **DOLIUM VARIEGATUM**, the upper figure (from Latin dolium—a jar with a wide mouth) is a yellowish brown shell, with dark brown spots, and exceeds six inches in length. Being fragile, and having a very wide mouth, perfect specimens are rare, although numbers of broken shells are from time to time washed up on the ocean beaches in the Province of Auckland. It lives in sand, but sometimes may be found crawling amongst rocks. It has no operculum. The Australian specimens are more handsome than the New Zealand ones. The Maori name is Pupuwaitai.

LOTORIUM RUBICUNDUM.—The lower figure, until lately known as the Triton nodiferus, from the old legend that it was the shell on which Triton blew at the bidding of Neptune to calm or rouse the waves, is a heavy, solid shell, varying a great deal in shape and colour; but usually brownish pink, variegated with dark brown. No difficulty will be found in identifying it. The specimens from Australia have more pink and less brown, and are not quite as fine as those of New Zealand. It is found on rocks and grassy banks in the North Island, but from being sluggish in its habits the point of the spire in large shells is usually worm-eaten, and good specimens over six inches long are seldom seen. The Lotorium tritonis, the largest univalve in the world, is similar to the Lotorium rubicundum, but not quite as solid or heavy. It has occasionally been found in the Northern part of New Zealand. The Maoris used it as a trumpet, fastening a mouth-piece to the spire. The Polynesian specimens of the Lotorium tritonis attain a length of nearly three feet, but nine or ten inches is the extreme length of our specimens. The Maori name is

Pupukakara, or Putara.

SIPHONALIA DILATATA (Plate IV.).—Fig. 1 has a pale yellow or greenish interior, the outside being reddish brown. Common on sandy, exposed beaches, and is sometimes over five inches long. The Maori name is Onare roa.

SIPHONALIA MANDARINA (Plate IV.).—Fig. 2 grows to the same length as the Dilatata; but is a narrower and more graceful shell. The interior is usually greenish. Found in the same localities as the Siphonalia dilatata.

SIPHONALIA NODOSA (Plate IV.).—Fig. 3 is a pretty shell, sometimes 2½ inches long. The interior is whitish, and the exterior the same colour, with purple and white markings. It is common on ocean beaches and sand banks in harbours.

STRUTHIOLARIA PAPULOSA (Plate IV.).—Fig. 4 is a handsome yellowish shell, with brown or purplish stripes. The interior is purple. The nodules on the whorls are very prominent. This shell is sometimes four inches long, and the lip, when the shell has attained full size, is remarkably strong and solid, forming a shell ring. From this it is known as the ring shell. In some places the lips, bleached to a perfect whiteness, come ashore in great numbers, the more delicate body of the shell having been broken to pieces among the rocks. These rings are sometimes seen strung together as ornaments. The lip does not form till the shell has attained its full growth, and though the shell is fairly common in the North Island, it is rare in the South. It is edible, and much esteemed by some people. The Maori name is Kaikai karoro, which is also the name for the Chione costata (Plate VIII., Fig. 26), and the Mactra æquilatera (Plate VIII., Fig 10). It is also called Tote rere.

STRUTHIOLARIA VERMIS (Plate IV.).—Fig. 5 is smaller than the Struthiolaria papulosa, which it resembles in its habits of growth. It is a pale brownish or yellowish shell, usually without nodules; and on the edge of each whorl nearest to the spire is a groove, as shown in the plate. The best Struthiolaria papulosa are found in the clean sandy margins of tidal channels, but their burrowing habits make them difficult to detect. I have never found the Struthiolaria vermis except cast up on ocean beaches, and it is comparatively rare. The Struthiolaria family, which derives its name from Struthio, an ostrich, as its mouth is supposed to be shaped like an ostrich's foot, is found only in New Zealand, Australia, and Kerguelen's Land. The Maori name is Takai.

EUTHRIA LINEATA (Plate IV.).—Fig. 6 (late Pisania lineata) is a solid, heavy shell, varying from grey to brown, and the lines shown in the plate are almost black. It is sometimes one and a-half inches long, and is found under stones and rocks. The colours vary very much, and the lines, in number and breadth, vary even more.

EUTHRIA FLAVESCENS (Plate IV.).—Fig. 9 (late Pisania flavescens) is a whitish or orange variety, with very pale markings, and much smaller than the Euthria lineata.

EUTHRIA VITTATA (Plate IV.).—Fig. 10 (late Pisania vittata) is a yellowish-brown shell, with broad brown bands. Another variety of the Euthria is somewhat like the Cominella lurida (Plate IV., Fig. 7) in shape and size. Another, the

Euthria littorinoides, is an orange-brown shell, but the interior of the aperture is a pale flesh-colour. In other respects, it is like the Euthria lineata. It is very difficult to draw any distinct line of demarcation between the varieties of this variable shell.

Figs. 7, 8, 11, 12, 13, and 14 are of the Cominella family, the New Zealand representatives of the voracious English whelks.

COMINELLA LURIDA (Plate IV.).—Fig. 7 is the most active and, for its size, the most voracious of our shellfish. Found in all harbours in the Province of Auckland, even up to high water mark, this greedy little animal, seldom more than an inch long, is well worth watching. In some localities, when a cart has been driven along a beach, the track, as soon as the tide reaches it, will swarm with the Cominella lurida. They are looking for cockles or other shellfish smashed by the wheel, and will even burrow in the sand to get at them. If you lift up a broken or injured cockle, some will cling to it with their rasp-like tongues till they are lifted out of the water. In calm, sunny weather, what looks like little bits of fat or candle-grease will be seen floating with the rising tide in very shallow water. These are Cominella lurida, which have perhaps eaten up everything in their vicinity, and have therefore decided to emigrate. A Cominella lurida, when shifting camp, will turn upside down, spread out its large white foot into a cup-shape, and let the rising tide sweep it along. They vary very much, from grey to purple or black, and sometimes even a mixture of two or more of these colours.

COMINELLA HUTTONI (Plate IV.).—Fig. 8 is a small pale brown shell, spotted with reddish-brown. The ridges on the exterior of the shell make it easy to identify.

COMINELLA MACULATA (Plate IV.).—Fig. 11 is a yellowish shell, with reddish-purple spots on the outside, the interior being also yellow. Its length is sometimes over two inches, and it is found in large numbers on sandy or shelly beaches, near low-water mark, in the North Island. Although a heavy, solid shell, it is of coarse texture, and therefore open to attacks by animal and vegetable parasites. A large specimen in good order is by no means common, the spire, or upper end of the shell, as shown in the plate, being usually worm-eaten.

COMINELLA TESTUDINEA (Plate IV.).—Fig. 12 is a handsome purple shell, the interior being darker than the exterior. It is about the same length as the Cominella maculata, but narrower, and the shell is thinner and harder. The exterior is covered with brown and white spots and splashes. It is common in the North Island and as far south as Banks' Peninsula. It is found on cockle banks and amongst rocks, especially those where sand is mixed with mud. The name Testudinea, from Latin testudo, a tortoise, is an appropriate one, as when held up to the light this Cominella looks like tortoise-shell.

COMINELLA VIRGATA (Plate IV.).—Fig. 13 is a greyish-brown shell, the raised lines, or ridges, that cross it being almost black. I have rarely found it, except amongst rocks in the harbours. It is much narrower than the Cominella testudinea, and not quite as long. The best way to obtain good specimens of these two Cominella is to break limpets, or other shellfish, and throw them into shallow water, close to rocks. In a few minutes, on revisiting the baits, the best specimens can be selected for the cabinet.

COMINELLA NASSOIDES (Plate IV.).—Fig. 14 is a pinkish-yellow shell, with very pronounced ridges on the exterior. The interior is brownish. So far, I have heard of its being found only in the South Island and the Chathams.

LOTORIUM OLEARIUM (Plate V.).—Fig. 1 (late Triton olearium) is a mottled brown and white shell, similar in its habits to the Lotorium rubicundum (Plate III.), but usually found on grassy banks in harbours at or below low water mark. The second figure on the plate is a good specimen of this shell, with its epidermis untouched, while the first figure has had the epidermis removed. To such shells as this and the Solenomya parkinsoni (Plate IX., Fig. 18) the epidermis adds an additional beauty, and to preserve it I have used a preparation of glycerine and chloride of calcium, being careful to put it on before the epidermis has time to dry or crack.

APOLLO ARGUS (Plate V.).—Fig. 2 (late Ranella argus) is a white or light grey shell, covered with a thin chestnut-brown epidermis. The lines that show so distinctly on the figure are dark chestnut. It is found on ocean beaches in both Islands, and attains a length of four inches.

APOLLO AUSTRALASIA (Plate V.).—Fig. 3 (late Ranella leucostoma) is a reddish-brown shell, covered with a fine hairy epidermis. The interior is purple. It is found amongst rocks in the open sea around the North Island. The edge of the lip is very deeply grooved. It attains a length of 4 inches.

LOTORIUM SPENGLERI (Plate V.).—Fig. 4 (late Triton spengleri) is a yellowish-white shell, covered with a pale brown transparent epidermis. The lines shown on the plate mark the grooves which cross the shell, and are slightly darker in shade than the ridges. It attains a length of five inches, and is found on the grass banks in sheltered places.

SEMI-CASSIS PYRUM (Plate V.).—Fig. 5, the helmet shell, from the Latin cassis, a helmet, is familiar to residents on the seaside, both in Australia and New Zealand, as it is a handsome shell, sometimes upwards of four inches in length. The colour varies a good deal, but is usually pinkish-white or pale chestnut, the wavy spots arranged in bands round the shell being usually dark brown. Sometimes the shell is nearly white. After heavy gales numbers are washed up on ocean beaches from the sandy banks on which they live.

SEMI-CASSIS LABIATA (Plate V.).—Fig. 6 (late Cassis achatina) is a smaller and narrower shell than the former, and somewhat rare. The dark markings are splashed, and not arranged in bands, thereby giving the shell a mottled appearance. The interior is brown or purplish.

LOTORIUM CORNUTUM (Plate V.).—Fig. 7 is a bright reddish-yellow shell, covered with a very long epidermis, which makes the shell appear more than double its real size. I have found a dozen or more of them on the ocean beaches in the Bay of Plenty. They were all dead shells, about one and a-half inches long, and the epidermis was wanting. The uneven, blunt-pointed lumps, with which this shell is covered, make it easily recognised. I have not heard of its being found anywhere in New Zealand, except in the Bay of Plenty, but it is fairly common in Sydney.

CALLIOSTOMA TIGRIS (Plate VI.).—Fig. 1 (late Zizyphinus tigris) is a whitish shell, striped or dotted in rows with red. Although sometimes over two inches across, the shell is thin and light. Its glistening interior, and shapely lines, make it one of our most handsome shells. These shells are sometimes found at low water mark, under and amongst rocks in harbours, as well as amongst kelp in the surf. When once a rock, or small patch of rocks, frequented by them is found, subsequent visits in the spring or early summer will nearly always be successful. It is common to both Islands. During the hot weather of summer, they apparently move to below low-water mark, and remain there in the deeper water until the winter. I obtained a considerable number of excellent specimens from a strip of rocks near the water tank at the entrance to Tauranga Harbour, but never found them except in spring or early summer. The Maori name is Mata-ngo-ngore, which name is also used for the Cantharidus family, on Plate VII.

CALLIOSTOMA SELECTUM (Plate VI.).—Fig. 2 (late Zizyphinus cunninghamii) is about the same width, but not the same height as the Tigris. The colour is white, with pale red spots arranged in rows around the spire.

CALLIOSTOMA PELLUCIDUM (Plate VI.).—Fig. 3 (late Zizyphinus selectus) is a whitish shell, covered with chestnut-coloured spots and splashes. It is about 1½ inches across.

CALLIOSTOMA PUNCTULATUM (Plate VI.).—Fig. 4 (late Zizyphinus punctulatus) is the commonest and least fragile of this family. It is seldom more than 1¼ inches across. Its rounded whorls, and prominent chestnut and white granules, make it easily distinguishable.

TROCHUS VIRIDIS (Plate VI.).—Fig. 5 is a greenish, cone-shaped shell. The interior is nacreous, and the exterior covered with coarse granules. The base, which is flat, is greyish. The figure but faintly shows the contour of this shell, which is a perfect cone. The young differ somewhat from the adult shells, and have a bright pink tip to the spire. In the plate the upper shell is a young one, and the two lower are adults. They are found amongst rocks at low water mark, in harbours, as well as in the surf. It is very difficult to extract the animal from the shell. Its maximum size is one inch across.

TROCHUS TIARATUS (Plate VI.).—Fig. 6 is usually white, with large grey or brownish-purple dots and bands on both the upper surface and the base, but it is a very variable shell. It is seldom as much as half an inch in length, and has a nacreous interior. It is covered with fine granules, and the base is flat. It appears to live slightly below low water mark, and can be easily obtained by dredging in harbours. The cup-shaped hollow at the base of the spire is much more pronounced than in the Viridis.

There is another not shown on the plate, the Trochus chathamensis, a small white shell, with pink or brownish-purple markings, that hitherto has only been found in the Chatham Islands.

ETHALIA ZELANDICA (Plate VI.).—Fig. 7 (late Rotella zelandica) is a well-polished, smooth shell, washed up in large numbers on the ocean beaches. The colours of the upper side vary, but are usually chestnut or purple waving lines on a yellowish-white ground. On the base is a circular band of purple round the columella, which is white. The interior is nacreous. Occasionally a shell is entirely pink, and then the circular band on the base is pink also. The largest shell I have seen was nearly one inch across, and, being very flat, would be only half an inch high. They appear to live in sandy ground, below low water mark in the ocean; and a dredge if drawn over one of their favourite spots will be filled with them. I have dredged half a bucketful at one cast between Karewa and Tauranga in five fathoms of water. The former name was Rotella zealandica, and Rotella, meaning a little wheel, well described the appearance of the shell, the waving line representing the spokes.

NATICA ZELANDICA (Plate VI.).—Fig. 8, a yellowish or reddish-brown shell, with chestnut-brown bands, the interior being pale brown, the mouth and its vicinity white. It is a clean, bright little shell, upwards of an inch across. Those in the ocean are lighter in colour, and larger and more solid than those found in harbours. As the tide falls in harbours, they conceal themselves near low water mark, especially in the vicinity of marine grass banks. When the tide is rising on a warm, sunny day, they spring out of the sand, dropping sometimes two or three inches from where they had been concealed. The operculum is horny, with a shelly outer layer; and the animal is prettily mottled and striped red and white.

There are two other Natica found in New Zealand, neither of which exceeds one-third of an inch across, and in shape are very like the N. zelandica. The Natica australis is a brown or grey shell, and the Natica vitrea is white.

NERITA NIGRA (Plate VI.).—Fig. 9 (late Nerita saturata) is a heavy, solid blue-black shell, with a whitish interior. This sombre-looking member of a handsome tropical family (of which the bleeding tooth Nerita is the best known) is sometimes over an inch in length, and found in large numbers clinging to the surf-beaten rocks of the North Island, quite up to high water mark. The operculum is shelly and prettily mottled with purple. This shell will stand boiling water, and, in fact, boiling water is required to kill the animal, which is quite as tenacious of life as an oyster. The Maori name is Mata ngarahu.

AMPHIBOLA CRENATA (Plate VI.).—Fig. 10 (the New Zealand winkle), lately known as Amphibola avellana, is an uneven, battered-looking shell of a mixed brown and purple colour, the interior being purple and the mouth whitish. It is an inch or more in length. Most mud flats up to high water mark are strewn with Amphibola. The natives eat this shellfish, which they call Titiko or Koriakai, in large numbers; but the muddy flavour, according to our ideas, makes it unpalatable.

MONODONTA SUBROSTRATA (Plate VI.).—Fig. 11 is a yellowish shell, about half an inch across, and is usually found near half-tide mark in harbours. The exterior is covered with black or bluish irregular bands. The interior is nacreous, and of a greenish colour, with a white patch round the columella.

MONODONTA AETHIOPS (Plate VI.).—Fig. 12 is a purplish-black shell, tesselated with white between the grooves. These grooves look like lines in the plate.

The interior of the mouth is white. Besides being usually covered with vegetable growth, part of which is seen in the illustration, the point of the spire is frequently worm-eaten and defective. This is the usual state in which all shellfish that herd together are found. It is upwards of an inch across, and found in large numbers amongst rocks, especially at the entrance to harbours, and from half-tide mark downwards.

MONODONTA NIGERRIMA (Plate VI.).—Fig. 13 has a smooth, purplish-black exterior, sometimes with small blue spots. The interior is white, and the shell about half an inch across.

MONODONTA LUGUBRIS (Plate VI.).—Fig. 14 is a thick, solid black shell, sometimes over half an inch across, and covered with coarse, irregular granules. The interior is white. This shell is found in large numbers under stones, at the entrances to harbours and sheltered beaches, almost up to high water mark.

There are six or seven other Monodonta in New Zealand, but they are small, and the four above described are the ones most likely to be met with.

TURBO GRANOSUS (Plate VI.).—Fig. 15 is a reddish-purple shell, varied with white, and is sometimes over 2½ inches across. The specimen photographed was much below the average size. The exterior is covered with well-defined rows of granules, while the interior is iridescent. It is found on rocks in the open sea in both Islands, but is a rare shell. The operculum is white and shelly.

TURBO HELICINUS (Plate VI.).—Figs. 16 and 17 (late Turbo smaragdus) is a blackish-green shell, found in great numbers at half tide mark on rocks all over New Zealand, especially at the entrance to harbours and in sheltered bays. Some are as much as 2½ inches across. The inside is white and glistening. The operculum is a solid, round, shelly one, with a greenish centre. In some specimens the outer side of the whorl, instead of being round and smooth, has two or three prominent raised ribs or bands on it. This variety is called Tricostata, and is represented by Fig. 16. I am inclined to believe it is only the young form of the ordinary variety. The Maori name is Ata marama.

ASTRALIUM SULCATUM (Plate VI.).—Fig. 18 (late Cookia sulcata) is a pinkish-brown shell, sometimes over 3½ inches wide. The interior is pearly, and the operculum is shelly, solid, and white. The laminae which cover the shell are easily bleached off, and when the shell is cleaned it has a handsome appearance. It is found in considerable numbers at low water mark amongst rocks on exposed beaches all over the North Island. The Maori name is Kakara or Kaeo, both of which names are also given to the Purpura haustrum (Plate II., Fig. 11).

ASTRALIUM HELIOTROPIUM (Plate VI.).—Fig. 19 is generally known as the circular-saw shell, and, although found all over New Zealand, is comparatively rare. It is reddish-purple, with an iridescent interior, and is sometimes over four inches in width. The shells on the plate are adults. The spines of the younger shells are much longer than those of adults. The best specimens have been dredged by oyster boats.

Plate VII.—Figs. 1 and 2 are Janthina, or violet shells, representatives of which are found all over the warmer parts of the world. The Janthinae live in great numbers on the surface of the ocean, being unable to sink, and are swept by gales and currents in every direction. At intervals, after very heavy gales, they come ashore in the Northern part of New Zealand in cart-loads; but after any ordinary gale a few specimens can be procured amongst the grass cast up by the highest waves. The animal, when touched, emits a quantity of violet-coloured fluid, the same colour as the shell. The shells are very light and fragile. A singular provision for its eggs is found attached to the female Janthina, in the shape of a float, or raft, to the under surface of which the eggs in little bags or capsules are attached, and there they remain until hatched.

JANTHINA EXIGUA (Plate VII.).—Fig. 1 is the smallest of the Janthina found in New Zealand, being rarely half an inch in width. The whorls are more rounded than in the other two varieties, and the spire is usually the same violet colour as the mouth, and the grooves on the shell are deep and prominent.

JANTHINA FRAGILIS (Plate VII.).—Fig. 2 is sometimes over an inch in width, the spire being much lighter in colour than the rest of the shell, frequently indeed being white. The grooves on the shell are fine, but clearly visible.

There is another variety occasionally found in New Zealand, the Janthina globosa, like the Janthina exigua in shape, but larger, and the grooving being very faint the shell has a glistening appearance. This variety is rare.

CANTHARIDUS IRIS (Plate VII.).—Fig. 3, from Iris, a rainbow, well describes the colour of this pretty little shell, seldom more than one and a-half inches in length. Pink, purple, yellow, and red seem to be the prevailing colours; and they are arranged in irregular waving lines on its smooth and polished surface. The interior is highly iridescent. It lives amongst seaweed and rocks below low water mark. The Maori name is Mata-ngo-ngore, which is also used for the Calliostoma shells on Plate VI.

CANTHARIDUS TENEBROSUS, var. Huttoni (Plate VII.).—Fig. 7 is a little bluish-black shell, about a-third of an inch long, with fine striæ or grooves running down the whorls. Alive, it is found in great numbers at low water on marine grass banks in harbours, and seems to be very active, as the anchors and cables of boats, moored for a few hours over one of their favourite haunts, will be liberally sprinkled with them.

CANTHARIDUS PURPURATUS (Plate VII.).—Fig. 8 is a heavier and rougher shell than the Iris, and of a rose-pink colour. Sometimes the whole shell is of this colour, but frequently only the top of the spire. It also lives amongst seaweed and rocks; but when living on grassy banks in harbours seems to lose its pink colour and become a pale grey.

CANTHARIDUS FASCIATUS (Plate VII.).—Fig. 9 (lately known as Bankivia varians) is found in Westland. White, green, rose, purple, or black in colour and plain or banded, and sometimes even with longitudinal wavy lines. It is about half an inch in length.

All of the Cantharidus family have beautiful nacreous interiors, and are the favourite New Zealand shells for necklaces and bracelets. When cleaned with acid, they are much admired. We have six or eight other varieties of Cantharidus, but they are small, and are not figured on the plate.

TARON DUBIUS (Plate VII.).—Fig. 4 is a shell about three-quarters of an inch long, and found under rocks in partly-sheltered harbours. The exterior varies from chocolate to black. The interior varies between purple and white. The lip end of the spire is usually reddish.

LITORINA CINCTA (Plate VII.).—Fig. 5 is a semi-amphibious shellfish common to both Islands. It is found amongst rocks in the open sea near high water mark. The exterior is brown, or bluish-black, with fine grooves or lines round it. The interior is violet, and the extreme length about ¾ inch.

LITORINA MAURITIANA (Plate VII.).—Fig. 6 is a very common shell in the North Island, where it is found on rocks in the open sea, or in harbours up to, and even above, high water mark. The shell is under half an inch long, and usually not more than a quarter of an inch. The colour outside is bluish-white, with a broad spiral band of dark blue. The interior is violet.

DAPHNELLA LYMNEIFORMIS (Plate VII.).—Fig. 10 is a very thin, whitish shell, with irregular brown markings, and is often dredged up in the vicinity of Auckland. Its extreme length appears to be 1¼ inches.

SURCULA NOVÆ-ZELANDIÆ (Plate VII.).—Figs. 11 and 12 (late Drillia zelandica) is a pale rose-coloured shell, nearly 1½ inches in length. It belongs to the Pleurotoma family, any of which can easily be identified by the notch in the outer lip, as shown near the centre of the figure. All of this family live below low water mark, and are obtained by dredging. It is found in both Islands.

SURCULA CHEESEMANI (Plate VII.)—Figs. 15 and 16 (late Pleurotoma) is a shell varying from pale pink to brown in colour. Interior rose or purple. The spire end is usually smooth. It is found in Auckland, and is about one inch in length.

SOLIDULA ALBA (Plate VII.).—Fig. 14 (late Buccinulus kirki) is a whitish shell, found in the North of Auckland. Its extreme length is ¾ inch.

POTAMIDES SUB-CARINATUS (Plate VII.).—Fig. 13 (late Cerithidea sub-carinata) is a dull black shell seldom over half an inch long. The colour is usually concealed by the reddish-brown epidermis. The interior is dark purple.

POTAMIDES BICARINATUS (Plate VII.).—Fig. 19 (late Cerithidea bicarinatus) is a reddish-brown or purple shell, covered with a blue or brown epidermis. The interior is purple. It is found in the North Island in large numbers on banks of sand mixed with mud near high water mark. Its extreme length is one inch.

SCALARIA ZELEBORI (Plate VII.).—Fig. 17 is the New Zealand representative of the Wentletrap family. It is a pure white shell, sometimes over an inch in length. The numerous ribs across the whorls are very prominent, and look like the steps of a ladder, whence it derives its name. It lives in the ocean below low water mark, and I have dredged it up with the Ethalia zelandica (Plate VI., Fig. 7). The Maori name is Totoro.

SCALARIA TENELLA (Plate VII.).—Fig. 18 is a dirty yellow, almost transparent, shell about a-third of an inch long. There is usually a pale brown band near the centre of the whorl. Found about half-tide mark in sheltered water.

TEREBRA TRISTIS (Plate VII.).—Fig. 20 is a bluish or blue-grey shell, slightly over half an inch in length. The interior is brownish-white, with a yellow band in the centre of the whorl. The varices on the exterior are not so prominent as in the Potamides (Fig. 13).

TENAGODES WELDII (Plate VII.).—Fig. 21 (late Siliquaria australis) is a small white shell, not more than one inch long. It is found in Hauraki Gulf.

TROPHON DUODECIMUS (Plate VII.).—Fig. 22 (late Kalydon duodecimus) is a pale yellow shell, usually covered with a thick, rough grey or brown coralline growth. The length is under half an inch; and it is found in the North Island amongst rocks on partly-sheltered beaches.

TROPHON PLEBEIUS (Plate VII.).—Fig. 23 (late Kalydon plebeius) is a brown or slate-coloured shell half an inch in length. The interior is reddish-purple, with six or eight narrow darker lines on the whorl.

TRICOTROPIS INORNATA (Plate VII.).—Fig. 24 is a pale brown or white shell, under half an inch in length, and found all over New Zealand.

MARINULA FILHOLI (Plate VII.).—Fig. 25 is a pale chestnut-coloured shell, with two large and one small white plaits on the inner lip. It is about a-third of an inch long, and is found in Auckland and Massacre Bay.

TRALIA AUSTRALIS (Plate VII.).—Fig. 26 (late Ophicardelus costellaris) is a brown, horny-looking shell, over half an inch long. It has two plaits on the inner lip. It is found in Auckland amongst mangroves near high water mark, and is also found in Australia. The maturer shells have narrow, dark brown bands on them.

TURRITELLA VITTATA (Plate VII.).—Fig. 27 is a yellowish-white shell, with spiral brown bands. It is under two inches in length, and found in the North Island.

TURRITELLA ROSEA (Plate VII.).—Fig. 28 is a reddish-brown, or yellowish, shell, finely banded with purplish-brown. It is found over three inches in length, and, though common enough in the North Island, is rare in the South. It is found amongst grassy banks during very low tides, point down, and almost buried in the sand. A sand bank of considerable size near Rangiawahia, in Tauranga Harbour, was inhabited by nothing but Turritella rosea. Four other kinds of Turritellæ are found in New Zealand, all smaller, but similar to the above.

TRIVIA AUSTRALIS (Plate VII.).—Fig. 29 is the New Zealand Cowry shell. It is less than ½ inch in length, and is white, with one or more flesh-coloured spots. It is found in the Northern part of Auckland Province and in Australia.

CYLICHNA STRIATA (Plate VII.).—Fig. 30 is a small, very narrow, smooth white shell. It is found in Auckland.

HAMINEA ZELANDIÆ (Plate VII.).—Fig. 31 is an exceedingly thin, horny, white or grey shell. It is sometimes called the sea snail, and is found on the marine grass in harbours, as well as in the open sea. Stray ones may be found in mud or sand.

BULLA QUOYI (Plate VII.).—Fig. 32 is a smooth, greenish shell, an inch and a-half long. It is sometimes marbled with purplish-grey, or with white dots. This shell is found in Auckland and Australia. The Maori name is Pupu wharoa.

BARNEA SIMILIS (Plate VIII.).—Fig. 1 is a white rock borer, up to two and a-half inches long. It is found all over the North Island, and at Waikowaiti, in the South Island.

PHOLADIDEA TRIDENS (Plate VIII.).—Fig. 2 is also a white rock borer, found up to nearly two inches in length. It seems particularly fond of the soft sandstone in the Auckland Harbour.

PANOPEA ZELANDICA (Plate VIII.).—Fig. 3 is a widely-gaping white shell, upwards of four inches long. It is common in the North Island, but rare in the South. It lives a considerable distance below the surface of the sand in the open sea or on exposed beaches. One, caught in situ, by Mr. C. Spencer at Cheltenham Beach (Auckland) was about eighteen inches below the surface of the sand at about half-tide mark. One species of the Panopea family, which is found in South Africa, lives at a depth of several feet. All bivalves that live in the sand have shells which gape more or less, apparently to enable them to push their syphons through the sand to the water. The deeper in the sand the shellfish lives, the longer and stronger the syphon must be. The Panopea burrows deeper than any other of our shellfish, and therefore requires the largest gape. As mentioned on page 6, bivalves do not leave their beds to feed, but push the syphon through the sand to the water and draw the water down one syphon and eject it through the other, absorbing the animal and vegetable matter as it passes through the mollusc's stomach. The Maori name is Hohehohe, which is also given to the Tellina family, on Plate VIII.

COCHLODESMA ANGASI (Plate VIII.).—Fig. 4 (late Anatina angasi) is a very white, almost transparent, thin shell, three and a-half inches long. One valve is nearly flat, and the shell gapes to a considerable extent at the narrower end. It is found in the open sea in sand in the North Island, Cook Strait, and Australia.

CORBULA ZELANDICA (Plate VIII.).—Fig. 5 is a yellowish or pinkish-white shell, with fine longitudinal lines on it. The interior is brownish, and the shell over half an inch long. It is common in the North Island and Australia.

SAXICAVA ARCTICA (Plate VIII.).—Fig. 6 is a rough, distorted, yellowish-grey shell, about three-quarters of an inch long. The interior is whitish. It is usually found in the roots of kelp or in sponges, and is obtained in both Islands.

MYODORA STRIATA (Plate VIII.).—Fig. 7 is a whitish or greyish-white shell, with fine longitudinal lines, the interior being pearly. It is 1¾ inches long. The right valve is rounded and the left valve flat. It is found in harbours, as well as on ocean beaches. The flat valves make excellent counters for card-players.

MYODORA BOLTONI (Plate VIII.).—Fig. 8 is a smaller and narrower shell than the Myodora striata, and the left valve is flat. In colour it is similar to the Striata. It is seldom over half an inch long, and lives on flat, sandy beaches. It is often found when sifting sand for small shells through a fine meshed sieve.

MACTRA DISCORS (Plate VIII.).—Fig. 9 is a large, rotund, greyish-white shell, with a blackish-brown epidermis. It is over 3½ inches across, and is found on sandy ocean beaches all over New Zealand. The Maori name is Kuhakuha.

MACTRA ÆQUILATERA (Plate VIII.).—Fig. 10 is a yellowish or white shell. It generally has a bluish-purple patch round the hinge. It is found on ocean beaches, and is over two inches long. The Maori name is Kaikaikaroro, which is also used for the Struthiolaria (Plate IV.), and Chione costata (Plate VIII.).

STANDELLA OVATA (Plate VIII.).—Fig. 11 is a thin, brownish-white, and somewhat wrinkled, shell over three inches long. The edge of the shell, and sometimes the whole shell, is covered with a brownish epidermis, the interior being yellowish. This shell is found all over New Zealand on muddy beaches, and especially near mangrove bushes in Auckland Harbour.

STANDELLA ELONGATA (Plate VIII.).—Fig. 12 (late Hemimactra notata) is a solid, greyish-white shell, four inches long. It is covered with an epidermis of pale chestnut, sometimes with darker chestnut bands, dots and splashes. The interior of the shell is yellowish.

RESANIA LANCEOLATA (Plate VIII.).—Fig. 13 (lately known as Vanganella taylori) is a smooth, white shell, covered with a thin, pale chestnut epidermis, the interior being white. It is upwards of four and a-half inches in length. It inhabits sandy ocean beaches in both Islands of New Zealand.

ZENATIA ACINACES (Plate VIII.).—Fig. 14 is a greyish-yellow shell, four inches long, and covered with a brown epidermis. The interior is bluish-green, pearly, and iridescent. This shell also inhabits the sandy ocean beaches of both Islands.

PSAMMOBIA STANGERI (Plate VIII.).—Fig. 15 is a purplish-white shell, sometimes rayed with darker purple. The interior is pinkish-purple. Its length is 2½ inches, and the shell is found in both Islands on sandy ocean beaches. The natives call it Wahawaha.

PSAMMOBIA LINEOLATA (Plate VIII.).—Fig. 17 is a purplish-pink shell, with darker concentric bands. Its interior is reddish-purple. This shell, which is found in both Islands on open ocean beaches, attains a length of 2½ inches. The Maori name is Kuwharu, or Takarape.

SOLENOTELLINA NITIDA (Plate VIII.).—Fig. 16 (late Hiatula nitida) is a thin, almost transparent, purplish-white shell, covered with a smooth, polished, horny epidermis. The interior is much the same colour as the exterior. Its length is about two inches. It is found in both Islands on sandy banks in harbours, and on sandy ocean beaches, but those found in harbours have sometimes little or no colour. The Maori name is Pi-Pipi.

SOLENOTELLINA SPENCERI (Plate VIII.).—Fig. 18 is a thin, almost transparent, milky-white shell. The interior is white. It is very like the Tellina alba (Fig. 21) in colour and general appearance, but much narrower, and the posterior end is curved and comes to a finer point. Its length is about two inches. I have found over a dozen live specimens washed up on Buffalo Beach, in Mercury Bay.

TELLINA GLABRELLA (Plate VIII.).—Fig. 19 is a smooth white, or pale yellow, shell, 3 inches in length, with a thin brown epidermis on the outer edge. The interior is chalky white. It is found on ocean beaches, but is also common on cockle banks in harbours. It lives some inches below the surface. Dead shells are found in considerable numbers, but the live ones are rare. The Maoris call this shell Hohehohe or Ku waru or Peraro. The name Hohehohe is also given to the Panopea (Plate VIII., Fig. 3).

TELLINA DISCULUS (Plate VIII.).—Fig. 20 is a clean smooth yellowish-white shell, with a bright yellow centre, the interior being the same colour as the exterior. Its length is 1½ inches, and it is found only in the North Island.

TELLINA ALBA (Plate VIII.).—Fig. 21 is a very thin, flat, nearly transparent, glistening white shell, the interior being the same colour. Its length is 2½ inches, and it is found on sandy ocean beaches in both Islands. The native name for this shell is Hohehohe, which name is also used for the Tellina glabrella.

TELLINA STRANGEI (Plate VIII.).—Fig. 22 (late Tellina subovata) is a whitish shell, similar to the Tellina alba, but more globose. It is under an inch long.

MESODESMA VENTRICOSA (Plate VIII.).—Fig. 23 (late Paphia ventricosa) is an opaque white, solid, smooth shell, found in the North Island, especially on the ocean beach near Kaipara. It is one of the many useful food molluscs we have. In the Kaipara district the natives take horses and ploughs on to the beach, and plough up the Mesodesma ventricosa like potatoes. Under the native name of Toheroa, a factory at Dargaville preserves these bivalves in tins. The specimen photographed was only a half-grown shell. In the Bay of Plenty I have found this shell seven inches long and extremely solid and heavy, and I am inclined to think from the shape and structure of the valve that the Bay of Plenty Mesodesma is different from the Ventricosa; but I never secured a live one while in Tauranga.

MESODESMA NOVÆ-ZELANDIÆ (Plate VIII.)—Fig. 25 (late Paphia novæ-zelandiæ) is the common oval Pipi, or Kokota, of the Maoris. This whitish shell, covered with a thin, horny epidermis, is sometimes 2½ inches long. It is found in both the North and South Islands on sandy banks in harbours and in tidal rivers.

ATACTODEA SUBTRIANGULATA (Plate VIII.).—Fig. 24 (late Paphia spissa) is a white shell, found in considerable quantities on sandy ocean beaches at half-tide mark. When the tide is flowing it is a very common sight to see great numbers of these bivalves washed up by the surf from their beds, and it is very interesting to watch the speed with which they can bury themselves again. They attain a length of about two inches, and are known to the Maoris as Tuatua or Kahitua.

CHIONE COSTATA (Plate VIII.).—Fig 26 (late Venus costata) is a strong, solid white shell, with thick radiating ribs. The only live ones I have found were either washed up on ocean beaches, or inside schnappers. This fish appears very fond of the Chione costata, and swallows it without attempting to crack the shell.

It attains a length of about two inches, and the Maoris call it Kaikai karoro, which name is also given to the Struthiolaria papulosa (Plate IV.) and the Mactra æqui-latera (Plate VIII.).

CHIONE STUTCHBURYI (Plate VIII.).—Fig. 27 (late Venus stutchburyii) is the common round cockle, found in both North and South Islands. Although when found on clean sandy banks it is usually reddish-brown on the outside and bluish-white inside, it varies in colour if the sand contains an appreciable quantity of mud. It is called Anga or Huai or Pipi by the Maoris, and attains a length of two inches.

CHIONE OBLONGA (Plate VIII.).—Fig. 28 (late Venus oblonga) is a brown or brownish-white shell, with a white interior, and is rather larger and more solid than the Stutchburyii, besides being more oval.

ANAITIS YATEI (Plate VIII.).—Fig. 29 (late Chione yatei) is a pale yellowish or brown shell, with a purple or slate-coloured patch round the hinge. The ridges on the outside, especially on the young shells, are thin and very high. As the shell attains its full size these ridges wear down. The old shells become thick and heavy, and are over two inches in width. It is found on exposed or ocean beaches in the North Island, and rarely in the South. The Maoris call it Pukauri.

HALIOTIS IRIS (Plate IX.).—Fig. 1 is the Pawa or Papa of the Maori, and the Mutton fish of the colonist. The outside is brown, and the inside a dark metallic blue and green, with an iridescent play of yellow and other colours. It is found on rocks in the open sea or on exposed beaches, and is six or seven inches wide.

HALIOTIS RUGOSO-PLICATA (Plate IX.).—Fig. 2 is about half the size of the Haliotis iris, and is known to the Maoris as the Pawa-rore or Koro-riwha. The outside is pinkish-brown, the interior being pale and highly iridescent. It is usual-ly found with the Haliotis iris, but is not so common.

Another Haliotis, named the Virginea, is much smaller and thinner than either of the above. The interior of this is like that of the Haliotis rugoso-plicata, but the exterior is variegated, and dotted and splashed with every conceivable colour. It is rare, and usually found on the sheltered side of small islands in the open sea.

GLYCYMERIS LATICOSTATA (Plate IX.).—Fig. 3 (late Pectunculus laticosta-tus) is a very solid, reddish-brown shell, sometimes (especially in the immature shells) splashed with chestnut and white. The six or eight teeth near the hinge on both valves are of even size and shape. It is usually found cast up on ocean beach-es. The shell attains a length and breadth of three and a-half inches. The younger shells have ridges or ribs on the outside, but these wear off with age. The Maori name is Kuakua.

GLYCYMERIS STRIATULARIS (Plate IX.).—Fig. 4 (late Pectunculus striatu-laris) is a small brownish shell, irregularly marked with chestnut, red, or white. The interior is whitish and brown, the exterior being smooth, and the extreme length of the shell about an inch. The markings of the hinge and teeth are similar to those of the Glycymeris laticostata.

CARDITA AVICULINA (Plate IX.).—Fig. 5 (late Mytilicardia excavata) is an irregular-shaped white shell, with yellow, pink, or dirty brown markings. The longitudinal grooves on the outside are very rugged and deep. The shell is over an inch in length, and is found in both Islands and in Australia.

RHYNCHONELLA NIGRICANS (Plate IX.).—Fig. 6 is an irregular-shaped, ribbed, black or dark brown shell, the left valve being much more rounded than the other. It is found up to one and a-quarter inches in breadth in the South Island and in the Bay of Plenty.

TEREBRATELLA SANGUINEA (Plate IX.).—Fig. 7 (late Terebratella cruenta) is an orange-red, evenly ribbed, shell up to one and three-quarter inches in breadth, found in the South Island. The left valve in this shell is nearly flat.

TEREBRATELLA RUBICUNDA (not shown on plate) is a smooth, pink, or dark red shell, of the same shape, but only half the size, of the Telebratella sanguinea, and found in considerable numbers in both Islands amongst stones. It is particularly plentiful amongst the stones on Rangitoto Island, in Auckland Harbour.

MAGELLANIA LENTICULARIS (late Waldheimia lenticularis) is not shown in the plate, but is a large, smooth, red or brown shell, two inches long, similar in shape to the above. All the above four shells, namely, the Rhynchonella, Terebratella (2), and Magellania, belong to the Terebratula family, and the right valve is longer than the left, and there is a small round orifice at the hinge end for the foot of the animal. On account of the resemblance these shells bear to the old Roman lamp, they are known as Lamp shells.

LITHOPHAGO TRUNCATA (Plate IX.).—Fig. 8 (late Lithodomus truncatus) is a thin brown shell, covered with a black or dark brown epidermis. It is found in the North Island, and attains a length of over one and a-half inches. It is a rock borer, and can bore into very hard rock. I have seen a small one that had bored into a thick Glycymeris shell.

VENERUPIS REFLEXA (Plate IX.).—Fig. 9 is a very irregular-shaped greyish shell, with prominent ridges on the outside. The interior is yellow, with a large blackish-purple patch. It is sometimes an inch in length, and is found in both Islands in the sand or mud, amongst rocks.

VENERUPIS ELEGANS (Plate IX.).—Fig. 10 is a white shell, with a white interior, and up to one and a-half inches long. The ridges on one end are very prominent. This shell is found only in the North Island.

DIVARICELLA CUMINGI (Plate IX.).—Fig. 11 (late Lucina dentata) is a milk-white shell, sometimes 1¼ inches in length. The grooves or furrows on the outside bend in the centre to almost a right-angle, giving it a peculiarly beautiful appearance, and making it easily recognisable. Found in both Islands on ocean beaches and in harbours.

VENERICARDIA AUSTRALIS (Plate IX.).—Fig. 12 (late Cardita australis) is a pale brownish-white shell, with prominent ribs. Sometimes the outside is marked and splashed with reddish-brown. The interior is white, with pink or rose-coloured patches. The shell is about one and three-quarter inches wide. It is found in both Islands attached to kelp roots, which usually discolour one end of the shell. The Maori name is Purimu.

CHIONE CRASSA (Plate IX.).—Fig. 13 (late Venus mesodesma) is a white or brown shell, one inch in length. It is found in large numbers on ocean beaches after a gale. The markings on it vary very much, and consist of radiating bands, or zigzag lines, of brown or purple brown. The interior is white, with a violet band round the margin.

TAPES INTERMEDIA (Plate IX.).—Fig. 14 is a brown or yellowish-white shell, with a white or grey interior. The young shells are marked with brown wavy or zigzag lines. It is found in both Islands on ocean beaches and in harbours, being sometimes over two inches wide. It is known to the Maoris as Hakari.

DOSINIA AUSTRALIS (Plate IX.).—Fig. 15 is a pale, pinkish-brown shell, with a white interior, turning to violet round the margin. It is found on ocean beaches in both Islands, and attains a length of three inches. The Maoris call it Tupa or Tuangi haruru.

DOSINIA SUBROSEA (Plate IX.).—Fig. 16 is a smooth copy of the above. It is pale pinkish-white, and found up to two inches long in the same localities as Dosinia australis. The Maori name for this shell is Hakari, the same as for Tapes intermedia.

There is another species of Dosinea (not shown in plate), about one inch long and pure white, found in the North Island. It is called Dosinia lambata.

BARBATIA DECUSSATA (Plate IX.).—Fig. 17 (late Arca decussata) is an ir-regular-shaped, brown or yellowish shell, the interior being white, varied with brownish-purple. It is covered with a long, brown, hairy epidermis. It is found in both Islands on ocean beaches and under rocks, and is up to three inches in length.

SOLENOMYA PARKINSONI (Plate IX.).—Fig. 18 is a dark brown, delicate shell, rayed with paler brown. The interior is greyish. The shining, thick, chestnut and black epidermis, which covers this shell, cannot be mistaken. It is found in both Islands on sandy banks in harbours, and is up to two inches in length. When the mantle is spread out in shallow water, this shellfish looks like a pink and purple flower.

MODIOLARIA IMPACTA (Plate IX.).—Fig. 19 (late Crenella impacta) is a brown shell, frequently with a mixture of green near the edge. The centre is smooth, but both ends are ornamented with fine radiating ridges. The interior is highly iridescent. The shell attains a length of 1½ inches, and is found in both Islands, in seaweed or grass and under rocks, both in harbours and on ocean beaches. The Maori name is Korona.

LIMA BULLATA (Plate IX.).—Fig. 20 is a white shell, about one and a-half inches long, and found in the North Island. Both it and the Lima zelandica are rare shells.

LIMA ZELANDICA (Plate IX.).—Fig. 21 (lately known as Lima squamosa and recently renamed Lima lima) is a beautiful white shell, with eighteen ribs. The spikes on the ribs are sometimes tinted with brown. It is found at Whangaroa

North, and has also been dredged up at Stewart's Island. It attains a breadth of 2½ inches. Although Lima lima is the latest name given this shell, I trust the name of Lima zelandica given it by Sowerby will be adhered to. It is quite as silly to duplicate the names of the family, to describe a species, as to have a kind of horse known as "horse horse." Crepidula crepidula (Fig. 27) is a similar instance.

SUB-EMARGINULA INTERMEDIA (Plate IX.).—Fig. 22 (late Parmophorus intermedia) is a white limpet-like shell, covered with a thin brown epidermis. It is sometimes 1½ inches long, the animal being like a large yellow slug.

SCUTUM AMBIGUUM (Plate IX.).—Fig. 23 (late Parmophorus unguis) is a white shell, covered with a thin brown epidermis, and is sometimes over 2½ inches long. The animal is like a big black slug, and, in comparison with the size of the slug, the shell is very small. A slug the size of a man's fist would have a shell about an inch long. Most shell-hunters would pass by a Scutum abiguum, not thinking it had a shell embedded in its folds. The shell is found amongst rocks in sheltered places on ocean beaches.

SIPHONARIA OBLIQUATA (Plate IX.).—Fig. 24 is like a brown limpet, about one and three-quarter inches long. On the right side is the siphonal groove, which is much more clearly defined in the Siphonaria australis (Fig. 25). The shell is found in Dunedin.

SIPHONARIA AUSTRALIS (Plate IX.).—Fig. 25 is a brown or chestnut-coloured limpet, up to one inch in length. The siphonal groove can be seen on the upper side of the figure. The best specimens I have found were on the piles of Tauranga Wharf.

EMARGINULA STRIATULA (Plate IX.).—Fig. 26 is a whitish limpet, about an inch in length. The notch, or fissure, which is a peculiar feature of this shell, is seen on the end of the shell facing the Lima zelandica (Fig. 21).

CREPIDULA UNGUIFORMIS (Plate IX.).—Fig. 27 is a parasite shell, over an inch long, and found inside the lips of other shells. It is a thin, clear white shell, and is well named, from unguis, a finger-nail, which it much resembles. It varies in shape from nearly flat to semi-circular, according to the curve of the part of the shell on which it grows. The Crepidula shells are easily identified by the shelly internal appendage, or lamina, in which the body of the animal rests. From the peculiar effect of this lamina the Crepidula shell looks like a boat. This shell has recently been renamed Crepidula crepidula, a silly duplication, like Lima lima (Fig. 21). The Maori name for the Crepidula is the same as for a limpet, namely, Ngakahi or Ngakihi.

CREPIDULA ACULEATA (Plate IX.).—Fig. 28 (late Crepidula costata) is an oval-shaped white parasite shell, with purplish lines on the edge. It is a common shell in the North Island, and found on rocks and amongst roots of kelp, and on the outside of other shells, especially mussels. It varies in colour and shape, but is usually deeply ribbed, and attains a length of 1½ inches.

There is another species of the Crepidula, viz., Monoxyla, similar in shape to the Crepidula aculeata, but white and smooth, and much smaller.

CALYPTRÆA MACULATA (Plate IX.).—Fig. 29 (late Galerus zelandicus) is a circular shell, found on rocks or kelp, and sometimes is attached to other shells, especially mussels. It attains a width of 1½ inches, and is covered with a brown, hairy epidermis.

HIPPONYX AUSTRALIS (Plate IX.).—Fig. 30 is a limpet, which takes its name from its shape, being like a horse's foot. There was a colony of some hundreds of this Hipponyx under a flat rock, resting on other rocks, on the ocean side of Mount Maunganui, at the entrance to Tauranga Harbour. Although there were thousands of other rocks round it, I never found the Hipponyx except under the one rock I have mentioned, and as far as I know it has never been found alive in any other part of New Zealand.

DENTALIUM NANUM (Plate IX.).—Fig. 31 is like a miniature white tusk of an elephant. It is about 1½ inches long. It is really a limpet, which, having chosen mud and sand as its habitat, has adapted itself to its surroundings and become long and thin, instead of broad and flat, like the rock-loving limpet. It is found on the West Coast of Auckland Province, especially between Manukau and Raglan.

ACMÆA OCTORADIATA (Plate IX.).—Fig. 32 is one of the dozen Acmæa found in New Zealand. It is a very flat shell, and lives amongst rocks in the surf.

ACMÆA PILEOPSIS (Plate IX.).—Fig. 33 is a nearly round, smooth limpet, the outside being blackish, spotted with white, and the interior bluish, with a black margin. It is about an inch across.

Amongst the other ten Acmæa found in New Zealand the most noticeable is the Acmæa fragilis, a very delicate, thin, green shell, with narrow brown bands. There is a green ring in the interior of the shell. It is found under stones, and is about ½ inch across.

PATELLA RADIANS (Plate IX., Fig. 34), and **PATELLA STELLIFERA** (Fig. 35) are two representatives of the many species of beautiful limpets we have. The limpet family has not had the attention of our scientists which it merits. The shells vary so much that it is extremely difficult to classify them. In the attempt to do so, Patella radians has been subdivided into five sub-species, but even this division is not a success. We have few more beautiful or interesting shells than limpets. We have them of every shape, and from three inches in width down to microscopic specimens. The limpet resides on one spot, but moves about with the rising tide in search of the vegetation on which it lives. This it mows down with its long scythe-like tongue, and, when satisfied, it returns to rest in its favourite spot. Limpets have the reputation of being indigestible, if not poisonous, but this is due to the head not being removed before the mollusc is eaten. If the head be removed carefully, the tongue, or radula, which is usually the length of the shell itself, will come with it. The 2000 or so fine teeth found on the average limpet's tongue will quite account for the belief that the fish is poisonous, as great irritation must be caused by these sharp little teeth. The Patella stellifera is usually found in caves or sheltered places amongst rocks exposed to the ocean swell. It is always covered with a coraline growth, usually of a pinkish tint, which growth has to be removed before the markings can be seen. Stars of all shapes, regular and irregular, will be found on the spire of the Patella stellifera. There is a reputation yet to be made by the man who can classify our New Zealand limpets. The Maori name for the limpet is Ngakihi, or Ngakahi, which name is also used for the Crepidula family.

PECTEN MEDIUS (Plate X.).—Fig. 1 (late Pecten laticostatus) is the well-known scallop found among the grass banks in harbours as well as in the open sea. The shells are sometimes five or even six inches across, and of all conceivable colours and mixtures of colours. The valve shown in the plate is the flat valve, which looks like a fan. The other valve, which is rounded, makes a good substitute for a scoop. This Pecten, or scallop, is the most delicate of our edible shellfish, but is never seen in our markets. The animal moves by opening its shell, slowly swallowing a large quantity of water, and in a rapid manner ejecting it, thereby pushing the shell backwards. The Maori name is Tipa.

PECTEN CONVEXUS (Plate X.).—Fig. 2 is a much smaller shell than No. 1, and quite as brilliantly coloured. The valves are nearly equal in shape. It is found amongst rocks, but is usually dredged in comparatively shallow water.

PECTEN ZELANDIÆ (Plate X.).—Fig. 3 is a still smaller shell, and the most brilliantly coloured of our Pecten family. The valves are similar in shape, and covered with short spikes. It has only the one ear, or lug, at the hinge end, but sometimes a portion of the ear is found on the other side. This shell lives amongst rocks, or in sponges, or on the roots of kelp, in sheltered or fairly sheltered portions of open beaches. It is found attached to the rocks by a byssus, or beard.

PINNA ZELANDICA (Plate X.).—Fig. 4 is generally known as the Horse Mussel. It is usually found amongst the grass, about low water mark, on sandy beaches, especially those containing a proportion of mud. The natives call it Huru-roa or Kupa, and in some places it is a staple article of diet with them. This horse mussel is found in certain spots in great numbers, and is then useless for a cabinet. The collector should look for odd scattered specimens. As a rule, only about half an inch of the shell will be found protruding above the beach, in very shallow water, but in deep water more of the shell will protrude.

MYTILUS LATUS (Plate X.).—Fig. 5 is the ordinary mussel, with a green epidermis, and the part near the hinge is usually eroded, as shown in the plate. It grows to a considerable size in New Zealand, being sometimes 8 inches in length, and is found in enormous quantities in favoured localities on rocks or attached by its beard in clusters to old cockle and other shells on the banks. About twenty years ago hundreds of acres of banks between the town of Tauranga and the sea were in one season colonised by mussel spawn, and although the mussel was before that date a rare thing on these banks, yet after the colonisation the banks were simply a mass of mussels, and the water, being only from one to two fathoms deep at low spring tide, they were easily procurable. On the other hand, banks near Kati Kati Heads, that were covered a few years ago, are now without mussels. This is probably due to some disease breaking out through overcrowding. The Mytilus edulis (not shown on plate) is a purplish shell, of similar shape and habits to the above, but much smaller in size. The Maori name for a mussel is Kuku or Porope or Tore-tore or Kutai, and for the smaller mussels Kukupara or Purewa or Toriwai.

MYTILUS MAGELLANICUS (Plate X.).—Fig. 6 is a bluish mussel, with prominent ribs, as shown in the plate. The interior is white, and the shell is found up to three inches in length.

VOLSELLA AUSTRALIS (Plate X.).—Fig. 7 (late Modiola australis) is a rough-looking, uneven shell, of a pale chestnut colour. It usually has a hairy-looking growth near the edge, as shown in the plate. It is found up to four inches in length.

There are two other of the Volsella family in New Zealand, neither of which are illustrated. The Volsella fluviatilis, a shiny, black mussel, shaped like the Edulis, and about 1½ inches long, found in brackish water, is the most common. The inside is bluish-white, and purplish round the margin.

OSTREA ANGASI (Plate X.).—Fig. 8 is a mud oyster, of which those dredged at Stewart's Island are the largest we have. Fine specimens were found in Ohiwa Harbour prior to the Tarawera eruption of 1886, but the deposit from that eruption appears for the time being to have destroyed them. There must be some large banks of this oyster in the Bay of Plenty, judging by the number of dead shells washed up in places; but, although I many times used the dredge while in Tauranga, I never had the good fortune to find one of the banks. Cartloads of the shells were at times washed up on the beach between the town of Tauranga and the entrance to the harbour.

The best known oyster in New Zealand is the Auckland rock oyster, the Ostrea glomerata (not shown in the plate), which is familiar to all who visit the seashore in the North. The Maori name for the rock oyster is Tio, and for the mud oyster Tiopara.

PLACUNANOMIA ZELANDICA (Plate X.).—Fig. 9 is of the family known in England as the pepper and salt oyster. The lower valve is flat and has the large oval opening, shown in the plate, through which the foot of the animal protrudes and holds the shell on to the rock. The shell is thin and fragile, and is found in both Islands. Another shell of the same family, the Anomia walteri (not shown on plate), is found at the Bay of Islands, and is usually coloured bright yellow or orange.

MUREX RAMOSUS, the last figure, is the latest addition to our New Zealand marine shells, and is described with the others of the Murex family on Plate II., and on page 9.

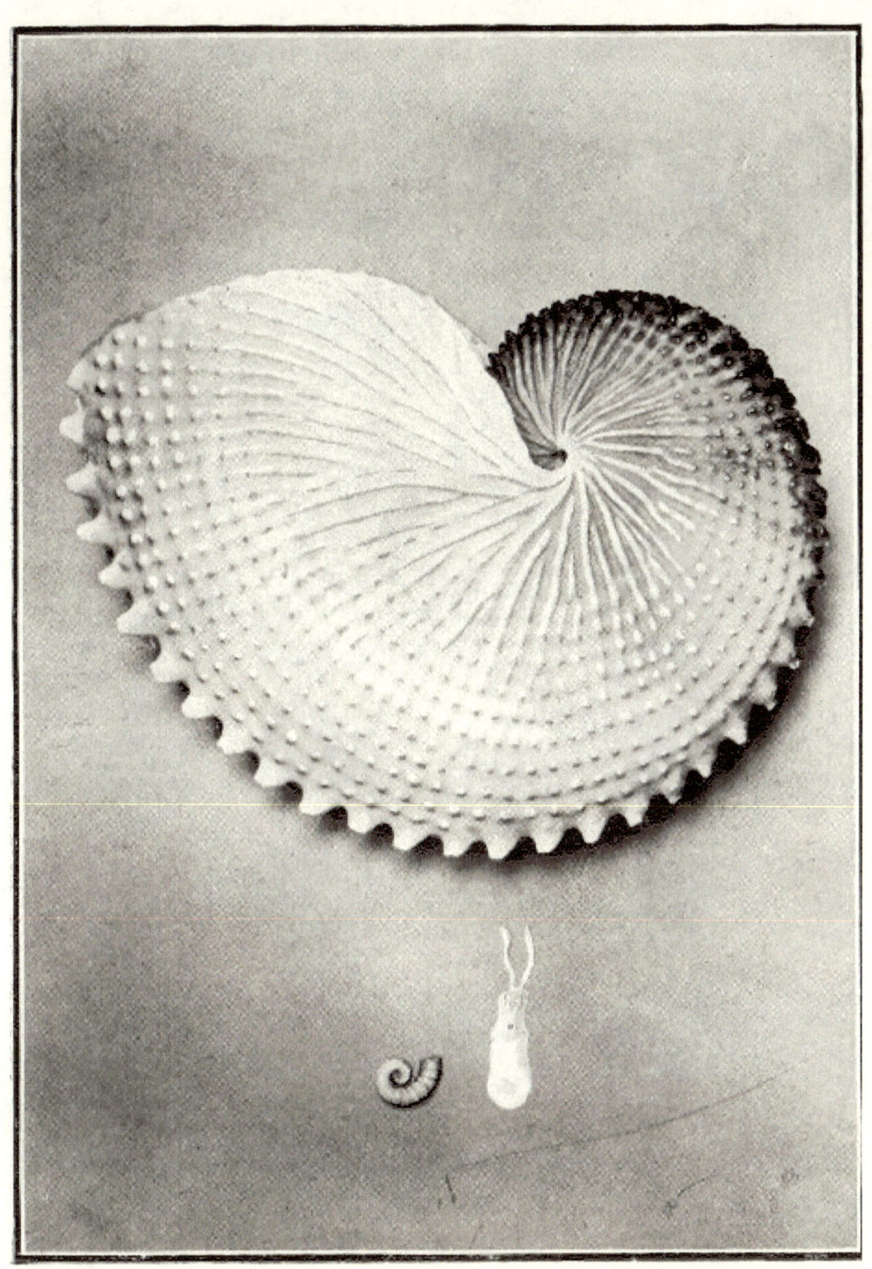

PLATE I.

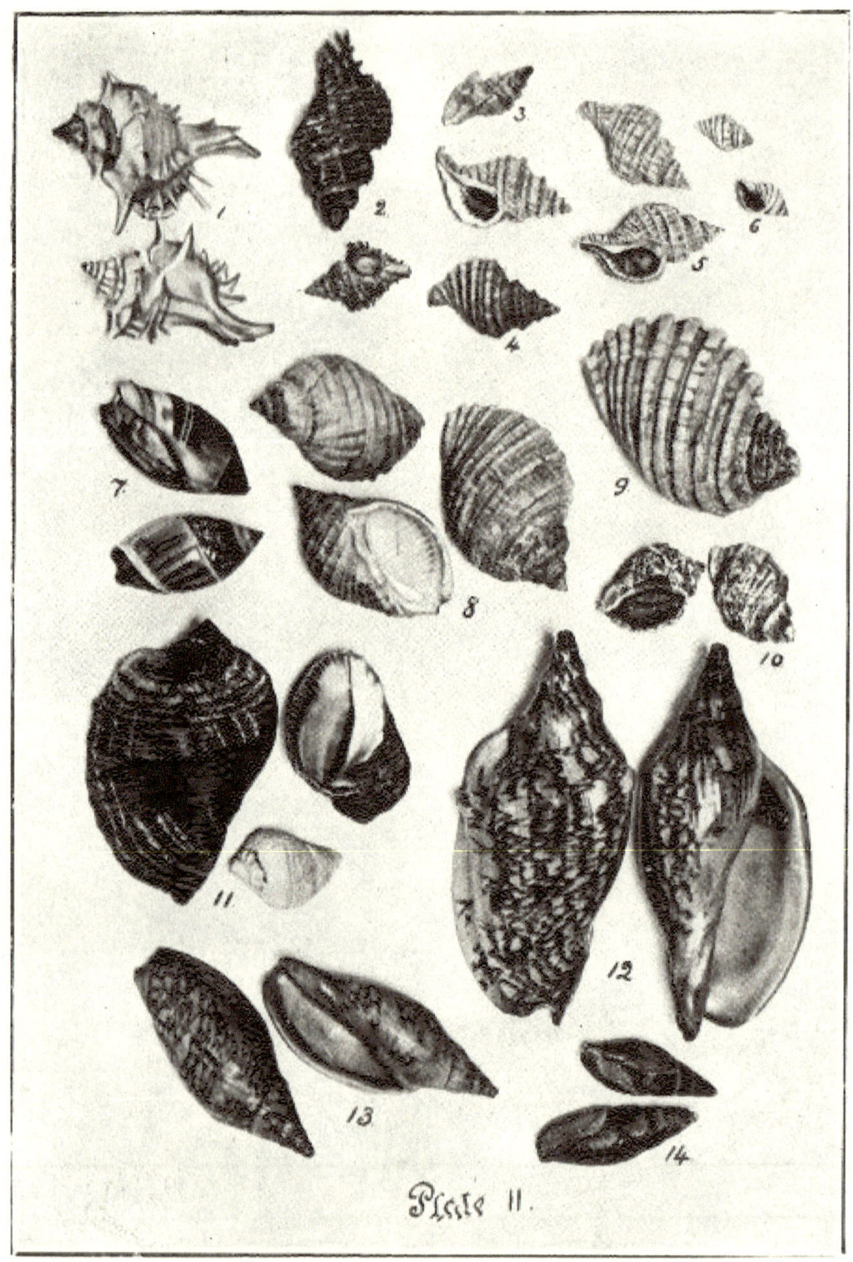

PLATE II.

		Page
1—	Murex zelandicus	9
2—	Murex octogonus	9

PLATE III.

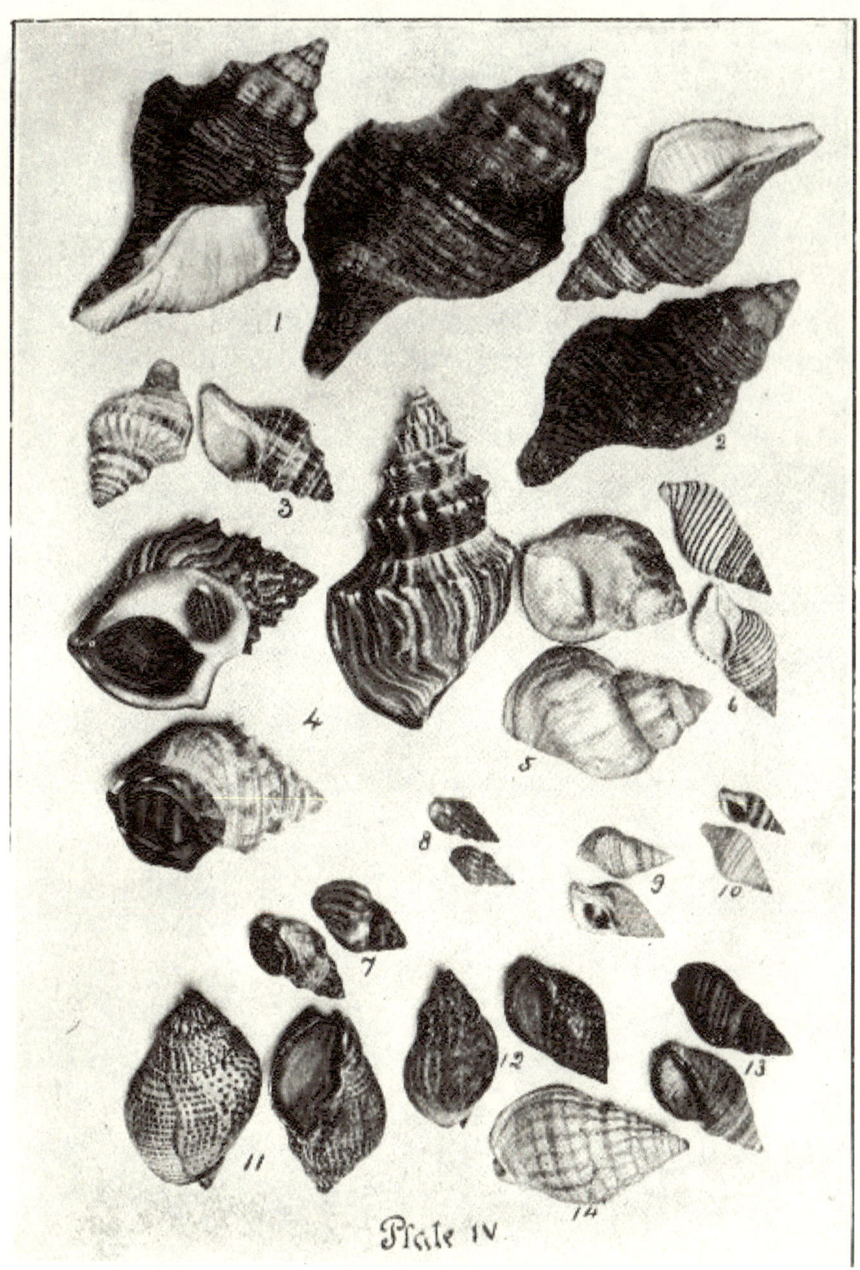

PLATE IV.

PLATE V.

<table>
<tr><td></td><td></td><td>Page</td></tr>
<tr><td>1—</td><td>Lotorium olearium</td><td>3, 14</td></tr>
<tr><td>2—</td><td>Apollo argus</td><td>14</td></tr>
</table>

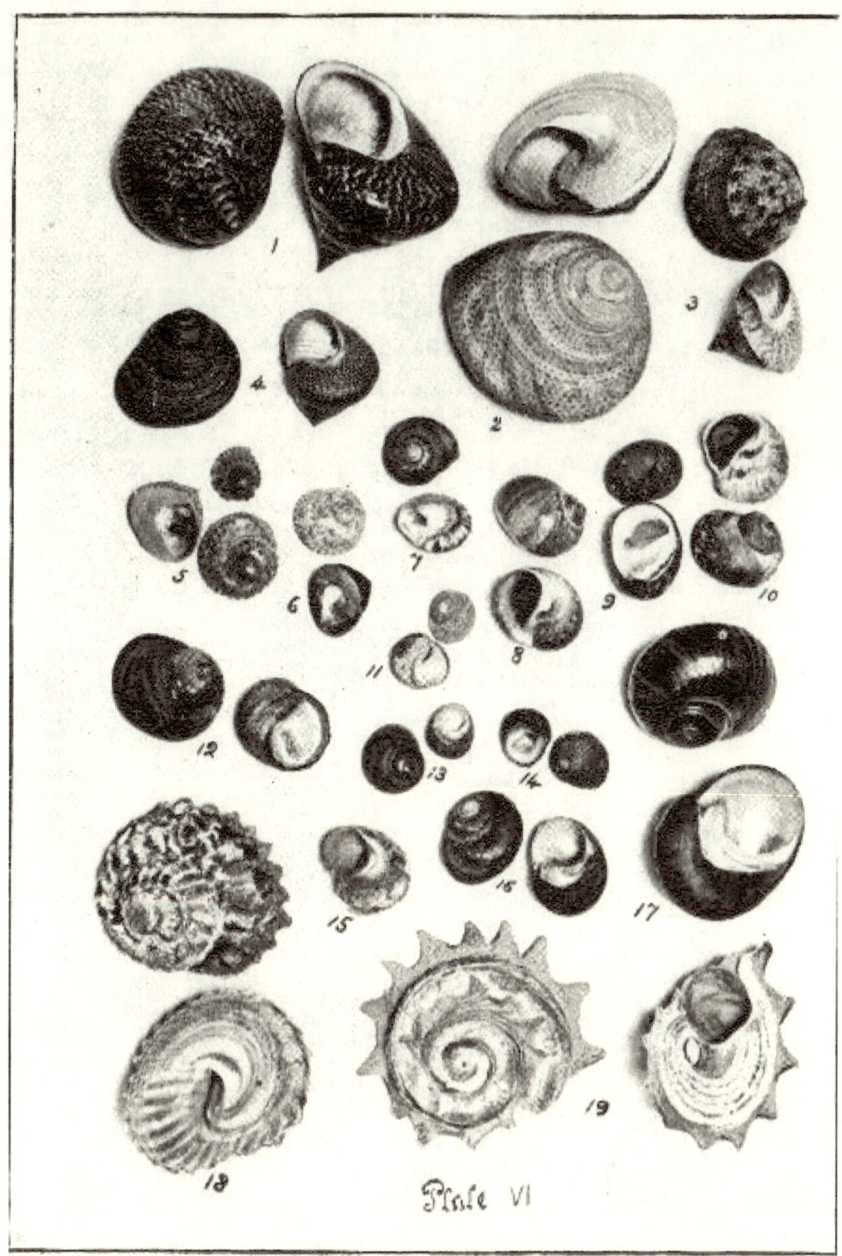

PLATE VI.

PLATE VII.

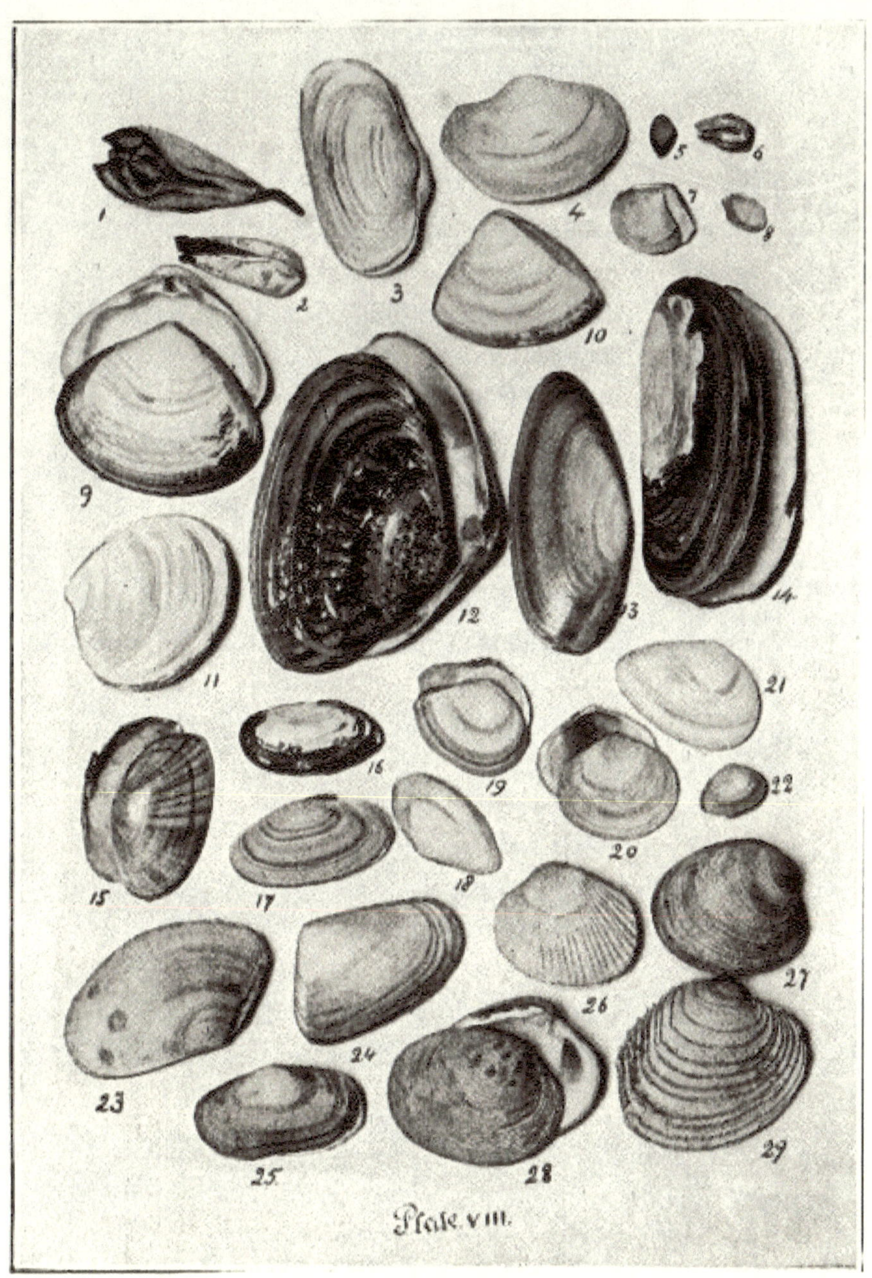

PLATE VIII.

PLATE IX.

<table>
<tr><td></td><td></td><td>Page</td></tr>
<tr><td>1—</td><td>Haliotis iris</td><td>24</td></tr>
<tr><td>2—</td><td>Haliotis rugoso-plicata</td><td>24</td></tr>
</table>

PLATE X.

INDEX.

D

E

G

H

Lector House believes that a society develops through a two-fold approach of continuous learning and adaptation, which is derived from the study of classic literary works spread across the historic timeline of literature records. Therefore, we aim at reviving, repairing and redeveloping all those inaccessible or damaged but historically as well as culturally important literature across subjects so that the future generations may have an opportunity to study and learn from past works to embark upon a journey of creating a better future.

This book is a result of an effort made by Lector House towards making a contribution to the preservation and repair of original ancient works which might hold historical significance to the approach of continuous learning across subjects.

HAPPY READING & LEARNING!

LECTOR HOUSE LLP
E-MAIL: lectorpublishing@gmail.com

9 789354 201837